Cesar David Torres Fernández
Armando Ibáñez Martínez
J. Refugio Tobar Reyes

LET'S TAKE CARE OF OUR PLANET ... SOWING LIFE

AF665258

Cesar David Torres Fernández
Armando Ibáñez Martínez
J. Refugio Tobar Reyes

LET'S TAKE CARE OF OUR PLANET ... SOWING LIFE

IT IS STILL TIME TO SAVE THE PLANET

ScienciaScripts

Imprint

Any brand names and product names mentioned in this book are subject to trademark, brand or patent protection and are trademarks or registered trademarks of their respective holders. The use of brand names, product names, common names, trade names, product descriptions etc. even without a particular marking in this work is in no way to be construed to mean that such names may be regarded as unrestricted in respect of trademark and brand protection legislation and could thus be used by anyone.

Cover image: www.ingimage.com

This book is a translation from the original published under ISBN 978-613-9-46721-1.

Publisher:
Sciencia Scripts
is a trademark of
Dodo Books Indian Ocean Ltd. and OmniScriptum S.R.L publishing group

120 High Road, East Finchley, London, N2 9ED, United Kingdom
Str. Armeneasca 28/1, office 1, Chisinau MD-2012, Republic of Moldova, Europe
Printed at: see last page
ISBN: 978-620-8-32694-4

Copyright © Cesar David Torres Fernández, Armando Ibáñez Martínez, J. Refugio Tobar Reyes
Copyright © 2024 Dodo Books Indian Ocean Ltd. and OmniScriptum S.R.L publishing group

LET'S TAKE CARE OF OUR PLANET SOWING LIFE

AUTHORS

César David Torres Fernández M.C. of the Benemérita Universidad Autónoma de Puebla Full-time professor-researcher in the economic-administrative area of the Facultad de Ciencias Agrícolas y Pecuarias. Tel. 01 (231) 31 22933 Cell. 231 387 6538 ghazzali.david@hotmail.com

Armando Ibáñez Martínez Dr. of the Benemérita Universidad Autónoma de Puebla Full-time research professor in the area of experimental designs and productive projects at the Faculty of Agricultural and Livestock Sciences. Tel. 01 (231) 31 22933 Cel. 231 1105539 armandoibama@outlook.com

Refugio Tobar Reyes DR. of the Benemérita Universidad Autónoma de Puebla Full-time research professor in the social-rural area, natural resources and feasibility projects at the Faculty of Agricultural and Livestock Sciences. Tel. 01 (231) 31 22933 Cel. 231 285 refugiotobar71@gmail.com

INDEX

Cuidemos nuestro
único hogar
TODA LA COMIDA
VIENE DE LA
TIERRA
¡CUIDÉMOSLA!
Naturaleza

I. AIM OF THE BOOK

The aim of disseminating this book is to raise awareness in various spheres of public and private life. The idea is to publicise the importance of reforestation as a major contribution to mitigating the problems of rural poverty and environmental degradation. In this way, its objectives are to rescue the countryside, reactivate the local economy and regenerate the social fabric in the communities.

We have done a lot of damage to our planet, we have deforested it indiscriminately, we have polluted the water, the land, the air and the fire we have used more to harm the planet than to benefit it. The good use of these four elements would go a long way towards having a sustainable and sustainable system throughout the length and breadth of our planet.

II. SOWING LIFE PROGRAMME

II.1 Target

Our objective is to contribute to the social wellbeing of agrarian subjects in their rural localities and to promote their effective participation in integral rural development. Mexico is a country rich in natural resources, biodiversity and culture.

The target population for the Sembrando Vida Programme is defined as those "Agrarian subjects of legal age who live in rural localities, whose municipalities are socially backward and who are owners or possessors of 2.5 hectares available to be worked in agroforestry projects.

II.2 Programme description

PROYECTO SEMBRANDO VIDA is a Federal Government programme that aims to combat rural poverty and environmental degradation by providing economic support to people who work in the countryside. The programme works to convert the ejidos into a strategic area for the full development of the Mexican countryside. In this way, its objectives are to rescue the countryside, reactivate the local economy and regenerate the social fabric in the communities, which is why it works in four components:

a). Productive inclusion

b). Care for the environment

c). Promoting a savings culture

d). Rebuilding the social fabric

How does Sembrando Vida work?

The federal government provides the plants and the resources to plant them. Each producer, community member and ejido member who joins this project will cultivate
2.5 hectares, for which they will receive a daily wage of 5,000 pesos per month. In addition, they will be part of a learning community made up of 25 farmers who will receive training accompanied by a social technician, who will work to promote the welfare of the community by fostering cooperative, harmonious and co-responsible relationships; a production technician, who will advise them on production techniques, soil types and the best adapted crops; three scholarship holders from the Youth Building the Future programme who live in the community and participate in the production and social processes.

Requirements

To be eligible for the programme you must be of legal age, live in rural localities, whose municipalities are in social backwardness and who are owners or holders of 2.5 hectares available to be worked in an agroforestry project.

Plot selection criteria

• Microaccount

• With potential for biodiversity recovery

• Degraded soils or with loss of forest vegetation cover - Productive reconversion, pastures, milpas and acahual management without graves

• Areas disturbed by disease, natural disasters or pests forestry

• *Plots where burning is practised are excluded.

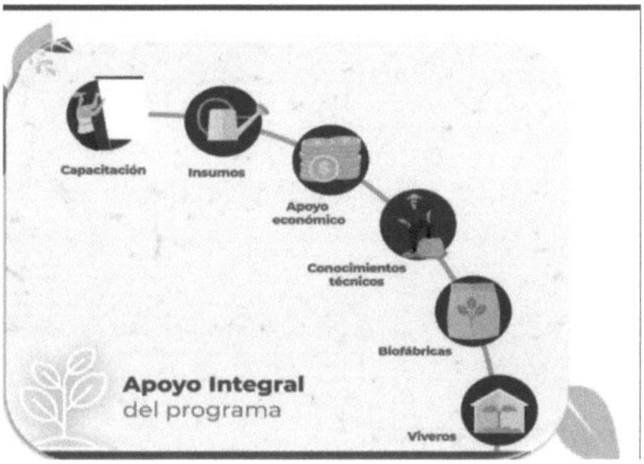

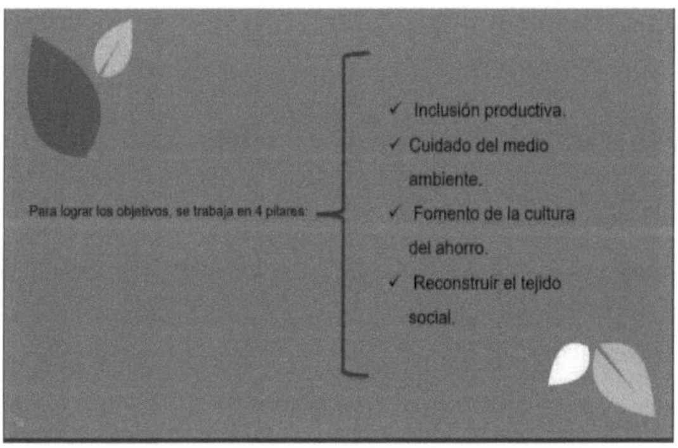

Alineación del Programa Sembrando Vida al Programa Sectorial de Bienestar

Objetivo Prioritario
Contribuir al bienestar social mediante ingresos suficientes, impulsar la autosuficiencia alimentaria, la reconstrucción del tejido social y generar la inclusión productiva de los campesinos en localidades rurales para hacer productiva la tierra.
Estrategias prioritarias
3.1 Instrumentar acompañamiento técnico agrícola con enfoque intercultural, apoyos económicos y en especie para que campesinas y campesinos puedan hacer productiva la tierra, lograr la autosuficiencia alimentaria y agroforestal, en coordinación con las instituciones públicas competentes.
3.2 Otorgar acompañamiento técnico social para promover la reconstrucción del tejido social en las localidades rurales.
3.3 Proporcionar asistencia técnica y mecanismos financieros con pertinencia cultural a campesinas y campesinos para promover su inclusión productiva y financiera.

Acciones puntuales	
3.1.1 Contratar y capacitar a técnicos agrícolas en temas de milpa intercalada entre árboles frutales, sistemas agroforestales, agricultura sustentable, biofábricas y viveros.	
3.1.2 Proporcionar asistencia y acompañamiento técnico a campesinas y campesinos con pertinencia cultural en temas de milpa intercalada entre árboles frutales, sistemas agroforestales, agricultura sustentable, biofábricas y viveros.	
3.1.3 Proporcionar apoyos económicos a campesinas y campesinos, con pertinencia cultural para que puedan hacer productiva la tierra.	
3.1.4 Proporcionar apoyos en especie a campesinas y campesinos con pertinencia cultural para que puedan hacer productiva la tierra.	
3.1.5 Apoyar la instalación de viveros comunitarios para la producción de planta.	
3.1.6 Apoyar la instalación de biofábricas que produzcan insumos para los viveros comunitarios y las parcelas de las campesinas y campesinos.	
3.2.1 Contratar y capacitar a técnicos sociales en temas de ahorro, construcción de ciudadanía, desarrollo comunitario, igualdad de género, salud comunitaria, derechos humanos, no discriminación y demás temas que aporten a la reconstrucción del tejido social en localidades rurales.	
3.2.2 Proporcionar asistencia y acompañamiento técnico social a campesinas y campesinos con pertinencia cultural en temas de ahorro, construcción de ciudadanía, desarrollo comunitario, igualdad de género, salud comunitaria, derechos humanos, no discriminación y demás temas que aporten a la reconstrucción del tejido social en localidades rurales.	
3.3.1 Contratar y capacitar a técnicos en temas de agroindustria, economía social y solidaria, encadenamientos productivos, canales de comercialización, incubación empresarial, organizaciones asociativas productivas y demás temas que promuevan la inclusión productiva de campesinas y campesinos.	
3.3.2 Proporcionar asistencia y acompañamiento técnico a campesinas y campesinos en temas de agroindustria, economía social y solidaria, encadenamientos productivos, canales de comercialización, incubación empresarial, organizaciones asociativas productivas y demás temas que promuevan la inclusión productiva de campesinos.	
3.3.3 Promover la constitución de un mecanismo de ahorro que apoye la inclusión productiva y financiera de los campesinos.	
Meta para el Bienestar	
3.1 Población con ingreso inferior a la línea de pobreza por ingreso en el ámbito rural	
Parámetro	3.2 Población con carencia de acceso a la alimentación en el ámbito rural

Types and Amounts of Support.

The support established in this Programme may be granted to the extent permitted by the 2021 Federal Expenditure Budget. The following support will be available to those entitled under the Programme:

1) Regular financial support, and where appropriate, additional support should be granted, deposited directly into their bank accounts, as set out in 3.6.1.

2) Support in kind as established in numeral 3.6.2, additionally, it will be able to count on social and technical accompaniment, through a process of training and permanent formation.

3) In the event of an act of God or force majeure, following a Declaration of Natural Disaster issued by the competent authority, the Technical Committee of the Programme Will may, subject to analysis and prior opinion, authorise extraordinary support in kind and/or financial support to compensate for the loss of plants and inputs. All of the above, for the correct and timely socio-community development, as well as agroforestry production.

ECONOMIC SUPPORT TO PROMOTE THE WELFARE OF THE SUBJECTS OF RIGHTS.

The mechanism for dispersing economic resources consists of the Programme making electronic transfers or issuing payment orders to the entitled parties.

• REGULAR FINANCIAL SUPPORT.

The person entitled, who has fulfilled his/her work program every month, will receive an economic support of $5,000.00 (Five thousand pesos 00/100 M.N.) of which $500.00 (Five hundred pesos 00/100 M.N.) will be destined as savings for the person entitled; of this amount, $250 (Two hundred and fifty pesos 00/100 should be destined to a savings investment in a financial institution, and $250 (Two hundred and fifty pesos 00/100 M.N.) should be destined to the Welfare Fund. Once a month, the social technician will inform the planters of the amount of their savings in the CACs verbally and/or in writing, which will be recorded in the corresponding Minutes, and each member must sign as informed, adding this document to the Minutes.

ADDITIONAL FINANCIAL SUPPORT

The Technical Committee of the Programme, after analysis and opinion, will determine those cases in which additional financial support may be given, under the following characteristics: The granting of additional financial support will only be determined by the Committee; the entitled parties may not apply for it directly. The Programme may allocate additional financial support to entitled persons for the purchase of seeds, vegetative material, plants and inputs.The entitled persons and/or the CACs may receive additional financial support, the amounts of which will be established by the Technical Committee of the Programme, and in accordance with the budgetary availability of the Sembrando Vida Programme.

IN-KIND SUPPORT FOR AGROFORESTRY PRODUCTION.

The mechanism for the delivery of in-kind support will be carried out as follows: all in-kind support will be delivered in each of the territories established by the Programme; the precise place of delivery will be determined by the regional and territorial coordinators, after informing the Undersecretary of Planning, Evaluation and Regional Development. Once the in-kind support is in the territories, the productive and social technicians will be responsible for coordinating the delivery to the entitled parties on the previously determined dates. For the granting of in-kind support, the Undersecretary of Planning, Evaluation and Regional Development may sign agreements, contracts, agreements and other necessary legal and administrative instruments with other entities, dependencies, etc., as well as with the Ministry of Agriculture, Livestock and Rural Development. and public and private institutions, in accordance with the regulations applicable in each case.

In-kind support

Plants, inputs, tools, community nurseries and biofactories In-kind support

The mechanism for granting support in kind will be carried out as follows: all support in kind will be delivered in each of the territories established by the programme; the precise place of delivery will be determined by the regional and territorial coordinators, after informing the Welfare Secretariat. Once the in-kind support is in the territories, the productive and social technicians will be responsible for delivering it to the agrarian subjects on the previously determined dates.

The agricultural producer will receive in kind:

Plants.

Plants required for planting. The quantity and type of plants, as well as the delivery and planting period, will be defined in the work plans.

Inputs.

The inputs necessary to develop the agroforestry programme in their production unit, which may vary according to the type of crop to be established in each territory.

Tools.

A package of tools to carry out the activities in your unit. production.
Toolkits will be considered according to the context and whether women and men are involved.

Community Nursery

Community nurseries will be established in each of the selected localities, which will have the necessary materials and inputs to produce 50,000 plants per year. The nurseries will be attended by producers with the accompaniment of the Productive Technicians. All producers, as long as they are part of the programme, will attend the community nurseries located in their territories, where the plants for their production units will be produced. If they leave the Programme, they will not have the right to the production in the nurseries, their participation is an integral part of the Programme and they will organise themselves territorially to dedicate the corresponding attention and participation time.

Biofactory.

Biofactories of inputs will be established in selected localities, which will have the necessary materials to produce bioferments, biopreparations and other agroecological substances that promote organic agriculture. The biofactories will be run by the producers themselves with the accompaniment of the Productive Technicians. The elaboration of compost will also be promoted, taking advantage of the material found in the production units. All of them, as long as they are part of the Programme, will attend the biofactories in their territories, where the organic inputs for their production units will be produced. In the same way, if any of them leave the Programme, they will not have the right to the production of the biofactories, their participation is a part of the Programme. The programme will be organised territorially in order to dedicate the corresponding attention and participation times.Liaison with the National Commission for Natural Protected Areas (CONANP).

According to Article 6 of the General Law of Social Development and Article 4 of the Political Constitution of the United Mexican States, referring to the General Law of Ecological Balance and Environmental Protection (LGEEPA), everyone has the right to a healthy environment for their development and well-being. The State shall guarantee respect for this right. Environmental damage and deterioration will generate responsibility for whoever causes it. According to Article 81 of the Regulation of the LGEEPA on Natural Protected Areas, natural resources may only be used if they are in accordance with the sustainable development schemes, the declaration of the natural protected area and its management programme. Therefore, the support granted by the Secretariat on territories within a federal protected natural area, or its zones of influence, must be validated by the competent body in the field of protected natural areas.

Challenging the suspension or cancellation of benefits

Persons whose benefits under the Programme are suspended or cancelled shall have the right to a Hearing Guarantee and, in addition, may challenge such determination before the Technical Committee of the Programme, by means of a free written document submitted in original and containing the following requirements:

- Name and address of the complainant.
- A section on the facts, which must be numbered and narrated in a clear and concise manner. in a clear and chronological manner.
- A section on evidence, in which mention is made of the public or private documents that relate to each fact indicated in the document, as well as the names of the persons to whom these documents relate.

• The paragraphs of these Operational Rules on which they are based

his brief.

• The signature of the petitioner. The time limit for lodging an appeal on the suspension or cancellation of benefits shall be 15 working days from the day following the day on which the contested decision is notified.

The Technical Committee of the Programme will resolve any challenges to the suspension or cancellation of benefits within a maximum of 95 days of receipt. In the event that any of the aforementioned requirements are missing, the Committee will ask the applicant to provide the missing documentation within a period of no more than 10 working days.

In the event of a successful challenge, the entitled person shall be entitled to reimbursement of the financial support for the months in which he/she did not receive the financial support.

In all matters not provided for in these Operating Rules, the procedure for suspending or cancelling the benefits of the Programme shall be governed by the provisions of the Federal Administrative Procedure Act.

III. IMPORTANCE AND BENEFITS OF REFORESTATION

Sembrando Vida has created more than 400,000 jobs in the countryside; for its thousands of planters and seeders it is no longer necessary to leave their places of origin in search of a livelihood, they now work in their own communities. The trees produce oxygen, purify the air, prevent the erosion, keep rivers clean, capture water for aquifers, serve as refuges for fauna, reduce soil temperature, favour the establishment of other species, regenerate soil nutrients, etc. Importance of reforestation Reverse soil erosion and revive watersheds. Forest trees, by slowing wind and falling water, protect the soil from erosion. Eroded and infertile soils damage agriculture and encourage landslides and flash floods. Impact of reforestation on the environment Reforestation, by planting trees in previously deforested or degraded areas, contributes to the capture and storage of CO_2. This helps to reduce the concentration of greenhouse gases in the atmosphere, mitigating climate change and its adverse effects.

Why is it important to reforest with native plants?

They allow us to recover spaces and biodiversity that have been lost through urbanisation.

How can we contribute to reforestation?

What to do before reforestation

- Know the site to be reforested. The site should be visited to identify the soil, vegetation and general conditions.
- Define the species. ...
- Weeding or removing weeds. ...
- Design plantation. ...
- Choose the right time of year.

What is conservation and reforestation?

Reforestation, the process of planting trees in areas where they have disappeared due to logging, urbanisation or land degradation, is an environmental conservation practice that goes far beyond landscape restoration and the promotion of biodiversity.

What is the importance of forests?

They generate oxygen. They control erosion, as well as the generation, conservation and recovery of soil. They contribute to carbon sequestration and the assimilation of various pollutants. They protect biodiversity, ecosystems and life forms.

What plants are used for reforestation?

Among those we know of that contribute to this process are the Pines, the Jiñocuagos, the Mangos, the Tatascames, the Liquidámbar and some Chilamates. The nectaries are the organs that secrete nectar, located in various places on the plant.

What is rural reforestation?

Rural reforestation is that which, in accordance with its objective, is established in forested or potentially forested areas where forests, jungles or semi-arid vegetation originally existed.

How do you succeed in reforestation?

The basis for successful reforestation is to use species appropriate to the site, germplasm of good quality, of known provenance and located within the immediate vicinity of the site to be reforested (Vázquez et al, 2001).

What are the benefits of reforestation?

Global reforestation offers multiple chemical, social and biological benefits. In addition to mitigating climate change, it protects biodiversity, improves air and water quality, and promotes human well-being.

Which trees purify the environment?

Melia, three-spined acacia, jacaranda and elm have the highest fixation capacity. Forest environment. Aleppo pine, stone pine and cork oak are the tree species that absorb the most CO_2.

The importance of reforestation in Mexico

Learn about these 5 factors that highlight the importance of reforestation and initiatives that help Mexico's ecosystems.

Did you know that Mexico is one of the most diverse countries in the world? However, it is losing its forests at an alarming rate per year, which is why the importance of reforestation must be emphasised more than ever.The loss of forest areas - whether by fire, pests or illegal logging - has several consequences. Primarily, CO_2 is no longer absorbed as much and the animals and plants of the local ecosystems are also at risk.

5 benefits of reforestation

A reforestation plan has multiple benefits for the environment and ecosystems, some of the most important of which are:

1. Reinforcing forest barriers

Trees are a natural barrier to cyclones and storms, as the wind loses its acceleration force when it hits the tree canopy. The same applies to rain and sea waves; a wooded area reduces the volume of water and filters it into the subsoil, helping to prevent flooding.

2. Regulates the climate

Through the shade they provide and transpiration, woodlands often help to dissipate high temperatures. This would be particularly useful in urban areas, where concrete and the number of buildings tend to accumulate heat.

3. Increases water availability

Trees also serve to channel rainwater, which will run down their canopies, leaves and trunks to the ground and hydrate it, or to people who want to collect water.

4. Prevents soil erosion

Another benefit of reforestation is that it prevents soil degradation. When it rains or there are landslides and there are no trees or vegetation to lessen the effect of these factors, the soil is more likely to erode.

5. Conserve biodiversity

The importance of reforestation lies in the fact that replanting trees and vegetation is a driving force that revives ecosystems. In this way, local wildlife is given a home and protection and, in turn, pollinators (such as bees) are attracted, which is very beneficial for nature.

Reforestation plan in Mexico

Broadly speaking, a reforestation plan attempts to return nature to its natural state by considering different actions to help achieve the objective.

- According to the National Forest Programme 2020-2024, Mexico has 20.8 million hectares that could potentially participate in the Sustainable Forest Management programme.
- In addition, as part of the Green Challenge 2021, the Ministry of Environment (SEDEMA) has already planted thousands of trees and shrubs in protected areas of Mexico City.

These are massive initiatives, but you can always ask for a reforestation plan in your area and help within your possibilities. Every action counts to help care for the environment, as well as to control and save on water consumption and electricity bills.

https://www.google.com/search?q=importancia+and+benefits+of+the+reforestation&rlz=1C1UUXU_enMX940MX940&oq=importance+and+benefits+of+the+reforestation&g

IV. DIFFICULTIES AND BENEFITS OF REFORESTATION

Reforestation

Reforestation, an alternative to reverse desertification Nature Desertification, or the loss of fertile and productive soil, is one of the problems that deepen the planet's climate crisis as the number of trees decreases and the greenhouse effect increases. One of the solutions is reforestation. Despite its disadvantages, it has become an alternative to turn thousands of hectares of land green again.

Reforestation is responsible for replanting areas affected by deforestation.

Without forests, life on Earth would not be possible. These, together with the oceans, are the lungs of the planet and their role in the fight against climate change is vital, absorbing some 2,000 million tonnes of CO2 - the main greenhouse gas and the main culprit in global warming - every year. The transcendence and value of these terrestrial ecosystems is so unquestionable that it is impossible to deny the importance of their role in the fight against climate change. that their care and respect is part of the Sustainable Development Goals (SDGs). Specifically, SDG 15: Life of terrestrial ecosystems seeks to protect, restore and promote their sustainable use. Almost a third of the planet is covered by forests, which is the equivalent of 4.06 billion hectares. These vast stands of trees provide us with the essentials for survival: the water we drink, the food we eat and the air we breathe. Yet we are destroying our source of life: 13 million hectares of forest are cleared by human hands every year. Overexploitation of natural resources, through logging or urban growth, is the main human-attributable cause of desertification, but there are others that are not human-induced. These include inconsistent rainfall and seasonal droughts, soil erosion and poor soils, or forest fires due to climate change. Against this backdrop, reforestation is one of the most effective strategies to reverse this problem.

What is reforestation. Objectives and benefits

Reforestation consists of repopulating deforested areas to recover forests destroyed in the recent past. Given the loss of large forest masses, which are vital for absorbing CO2, generating oxygen and combating climate change, it is necessary to plant new trees on a massive scale in order to prevent the loss of ecosystems and slow down the deterioration of the planet.

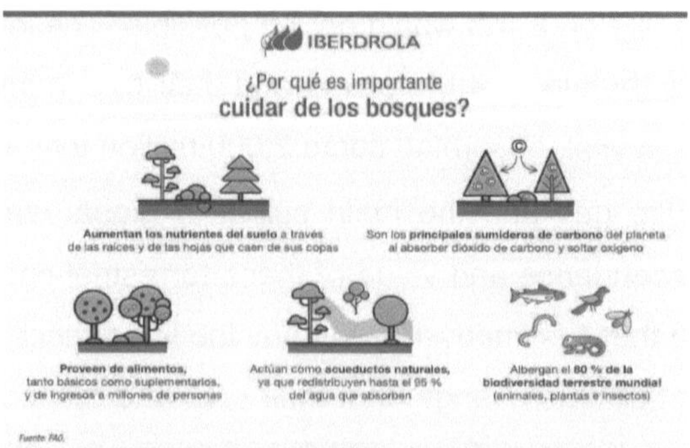

Reforestation therefore contributes to the achievement of several objectives, all of which are aimed at restoring the stability that forests provide:

Restoring biodiversity loss

Forests are home to more than 80% of all terrestrial species in the world. Specifically, according to The State of the World's Forests 2020 (FAO), they are home to more than 60,000 tree species, 80 % of amphibians, 75 % of birds and 68 % of mammals. Their degradation and disappearance is driving hundreds of species to extinction despite conservation efforts. The European Union Biodiversity Strategy 2030 foresees the planting at least 3 billion trees on European soil to contribute to the protection of biodiversity.

Reducing carbon dioxide in the air

Human activity causes annual emissions of around 40 Gt CO2. Half of these gases stay in the atmosphere contributing to global warming and the other half is absorbed by forests and oceans. Reforestation is essential for our own livelihoods: forests are essential carbon sinks for slowing climate change. Without them, the average global temperature will continue to rise, causing sea levels to rise and glaciers and poles to melt, among other climatic effects.

Reversing soil erosion and revitalising watersheds

Forest trees, by slowing wind and water flow, protect the soil from erosion. Eroded and infertile soils are detrimental to agriculture and favour landslides and flash floods. Reforestation aims to alleviate this situation, which is also accentuated by indiscriminate logging, by preserving soil fertility with well-attached roots. In turn, watersheds are revived by the recovery of nutrients.

Caring for human health

Deforestation and its effects on habitat not only deprive us of essential nutrients, but are also the main routes of transmission of emerging infectious diseases, including COVID-19. Seventy-five percent of these diseases, including avian influenza and Ebola, are transmitted from wildlife to humans. people. Degradation of forests, without optimal reforestation, leads to human exposure to zoonotic diseases.

Types of reforestation

Depending on the location, there are two types of reforestation:

- Urban reforestation. This refers to the planting of trees in urban environments. Its objective has to do with the city's own needs: to modify the climate -green spaces are good for combating heat-, to improve air quality -the high incidence of traffic in cities causes CO2 levels to rise-, to increase shaded areas or to beautify the environment.

- Rural reforestation. This is the massive planting of trees in forested areas that have been deforested, i.e. where forests, jungles or semi-arid vegetation used to exist. It can also occur in areas where these did not previously exist, although the appropriate term in this case would be afforestation. Within rural reforestation, there are different subtypes depending on the objective: conservation, protection and restoration, agroforestry or productive.

How reforestation is carried out

To reforest land that has been left fallow by indiscriminate logging, fire or deforestation due to climate change, a plan must be drawn up. Reforestation must be sustainable, i.e. it is not about planting trees arbitrarily. Here are the main issues to consider:

Conduct a field study

First of all, it is necessary to study the terrain and check the conditions of the site: from the soil - depth, texture, fertility - to the climate - dry or wet season (humidity is essential) - or the type of population living in the ecosystem - native fauna and flora.

Choosing the species for restocking

It is best to opt for indigenous species, but imported fast-growing species that are compatible with the soil and climate can also be included. The forest germplasm should be of good quality and the nursery should ideally be no more than 100 kilometres away. The manner and timing of transport is also important, avoiding the sun or strong wind currents.

Opting for a planting method

We must prepare the soil, choose the right tools and opt for the least invasive technique. In addition, we must take into account the height and cover of each new plant so that they do not damage each other. Planting does not end with the introduction of the forest germplasm, but a follow-up plan must be carried out.

Establish a protection plan

Within the monitoring plan, ways to protect the reforested forest from disease, pests, fire, illegal logging, etc. should be developed. Maintenance and evaluations are vital in order to secure the reforestation.

Problems of reforestation

According to the Global Forest Resources Assessment 2020 (FAO), more than 2 billion hectares of forest worldwide have management plans. The preparation of guidelines for the proper functioning of forests is a starting point that should not be missing when rehabilitating a forest or embarking on reforestation because, as in any process, problems can arise:

- If reforestation is impulsive, i.e. without a good implementation plan, it can be counterproductive, damaging species diversity or agricultural crops.

• In large forestry crops we can achieve the opposite effect to the one we are looking for, drying out and impoverishing the soil due to excess salt concentration.

• A poor choice of new trees to be introduced, as well as the way they are planted and positioned, can be detrimental. In addition, the introduction of invasive species can lead to the extinction of others.

• Poorly planned reforestation could lead to monoculture, affecting not only the diversity of the native flora but also the habitats of the different forest dwellers.

https://www.iberdrola.com/sostenibilidad/que-es-reforestation#:~:text=Problems%20of%20the%20reforestation%20reforestation%20and%20text=We%20can%20problem%20and%20positioning%20these%20can%20be%20damaging%20.

Problems of reforestation

In large forest plantations, we can achieve the opposite effect to that sought, drying out and impoverishing the soil by excess saline concentration. A poor choice of new trees to be introduced, as well as the way they are planted and positioned, can be detrimental.

What are the disadvantages of reforestation?

Reforestation of trees can worsen the problems it is intended to solve. Tree planting campaigns can combat climate change, improve communities and restore biodiversity. However, if done badly, it can accelerate extinction.

Does reforestation of trees help or harm the planet?

Reforestation of trees may worsen the problems it is intended to solve. By Catrin Einhorn. 21 Mar 2022 The New York Times

Tree planting campaigns can combat climate change, improve communities and restore biodiversity. However, if done badly, they can hasten extinction.

For every T-shirt purchased, a tree is planted; for every bottle of wine; for every time you swipe your credit card. Countries plant trees to meet their global commitments and companies plant trees to reinforce their sustainability record.

El Times. A weekly selection of Spanish-language stories you won't find anywhere else, with accents and ñ. Get it sent to your inbox.

As the climate crisis deepens, businesses and consumers are joining non-profit organisations and governments in a global reforestation boom. Last year, billions of trees were planted in dozens of countries around the world. These efforts can be a triple win, providing livelihoods, absorbing and locking in global warming carbon dioxide and improving ecosystem health.

But when projects are done badly, they can worsen the very problems they are intended to solve. Planting the wrong trees in the wrong place can reduce biodiversity and accelerate the extinction of ecosystems, making them much less resilient. Tackling biodiversity loss, already a global crisis similar to climate change, is becoming increasingly urgent. Extinction rates are increasing. An estimated one million species are at risk of disappearing, many of these within a matter of decades. And the collapse of ecosystems not only threatens animals and plants, but endangers the food and water supplies on which humans depend.

In the midst of this deepening crisis, companies and countries are increasingly investing in planting trees that cover large areas with non-native commercial species in the name of combating climate change. These trees absorb carbon, but do little to support the webs of life that once thrived in these areas. "In essence, it creates a barren landscape," said Paul Smith, who heads Botanic Gardens Conservation International, a group of organisations working to prevent plant extinction. "If people want to plant trees, let's make it positive for biodiversity as well.

In the world of reforestation there is a golden rule: plant "the right tree in the right place". Some add: "for the right reason". However, according to interviews with various stakeholders - scientists, policy experts, forestry companies and forest stakeholders' organisations - it is clear from the interviews that the reforestation - people often disagree about what "suitable" means. For some, it is about large nurseries for carbon storage and timber. For others, it is about providing fruit trees to small farmers. For others, it is about allowing regeneration of carbon and timber species.For others, it is about providing fruit trees to small farmers. For others, it is about allowing endemic species to regenerate.

According to restoration experts, the best efforts try to meet a range of needs, but it can be difficult to reconcile competing interests. "It's kind of like the Old West," said Forrest Fleischman, professor of politics. University of Minnesota. A natural solution The Earth does not have enough land to tackle climate change with trees alone, but in combination with drastic cuts in fossil fuels, trees can be an important natural solution. They absorb carbon dioxide through the pores of their leaves and store it in their branches and trunks (although trees also release carbon when they burn or rot). This ability to sequester CO2 is why forests are often called carbon sinks. In Central Africa, Total Energies, the French oil and gas giant, announced plans to plant trees on 40,000 hectares in the Republic of Congo. The project, located on the Batéké plateau, an undulating mosaic of grasslands and wooded savannahs with patches of denser forests, would sequester more than ten million tonnes of carbon dioxide over 20 years, according to the company. "Total is committed to developing natural carbon sinks in Africa," Nicolas Terraz, then Total's senior vice-president for Africa, exploration and production, said in a company press release about the project in 2021. "These activities build on the priority initiatives taken by the group to avoid and reduce emissions, in line with its ambition to reach carbon neutrality by 2050."

In order to achieve this goal, companies must remove at least as much carbon from the air as they release. Many, like Total Energies, are turning to trees to achieve this. On the Batéké plateau, a species of acacia from Australia, intended for selective logging, will cover a large area. The project, which is part of a Congolese government programme to expand forest cover and increase carbon storage, will create jobs, according to the company, and ultimately increase the biodiversity of the ecosystem by allowing local species to grow for decades.

But scientists warn that the plan may be an example of one of the worst kinds of afforestation efforts: planting trees where they would not occur naturally. Such projects can devastate biodiversity, threaten water supplies and even increase temperatures because, in some cases, the trees absorb heat that grasslands - or, in other parts of the world, snow -

would have reflected. "We don't want to do harm under the guise of doing good," said Bethanie Walder, executive director of the Society for Ecological Restoration, a global non-profit organisation.

The Batéké Plateau is one of the least studied ecosystems in Africa, according to Paula Nieto Quintano, an environmental scientist who has focused on the region. "Their importance for local livelihoods, their ecology and ecosystem functions are poorly understood," said Nieto.

Those who study forest conservation insist that trees are not a panacea.

"I'm afraid a lot of companies and governments see it as an easy way out," said Robin Chazdon, Professor of Rainforest Restoration at the University of the Sunshine Coast in Australia. "They don't have to work as hard to reduce their emissions because they can just say, 'Oh, we offset it by planting trees'."

Doing the right thing

All trees store carbon; however, their other benefits vary greatly depending on the species and where they are planted. For example, eucalyptus grows fast and straight, making it a lucrative timber product. Native to Australia and some northern islands, its leaves feed koalas, which have evolved to tolerate a potent poison they contain. But in Africa and South America - where the trees are grown for timber, fuel and, increasingly, carbon storage - they are of much less value to wildlife. They are also known to cause water depletion and worsening forest fires.

Experts recognise that forest restoration and carbon storage are complex, and that commercial species play an important role. People need wood, a renewable product with a lower carbon footprint than concrete or steel, as well as paper and cooking fuel. Sometimes planting fast-growing species for harvest can help conserve surrounding endemic forests. And, by strategically adding native species, nurseries can help biodiversity by creating wildlife corridors to link disconnected habitat areas.

"This restoration movement cannot happen without the private sector," said Michael Becker, communications officer for 1t.org, a group created by the World Economic Forum to drive the conservation and growth of a trillion trees with the help of private investment. "Historically, there have been bad actors, but we have to get them to come on board and do the right thing. One challenge is that supporting biodiversity does not offer the financial returns of carbon storage or timber markets. Many governments have set standards for reforestation, but they often offer a wide margin for manoeuvre.In Wales, one of the most deforested countries in Europe, the government offers incentives for reforestation. But those doing the work only have to include 25 per cent endemic species to qualify for government subsidies. In Kenya and Brazil, rows of eucalyptus trees grow on land that was once ecologically rich forest and savannah. In Peru, a company called Reforesta Peru is planting trees on degraded Amazonian land, but is increasingly using cloned eucalyptus and teak for export. Investors prefer them because they command better prices, says Enrique Toledo, general manager of Reforesta Peru. "They are very well known internationally and there is an unsatisfied demand for timber. When researchers from University College London and the University of Edinburgh assessed national commitments to reforestation and restoration, they found that 45 per cent involved "planting vast monoculture tree plantations as profitable enterprises".

Biodiversity balance

When companies pledge to plant a tree for every purchase of a particular product, they often do so through non-profit organisations that work with communities around the world. This support can take the form of reforestation following forest fires or providing fruit and nut trees to farmers. But even these projects can compromise biodiversity.

The planet is home to almost 60,000 tree species. According to a recent report, one third of them are threatened with extinction, mainly due to agriculture, grazing and exploitation. But only a fraction of the species are planted worldwide, according to tree-planting groups and scientists.

"The same species are planted all over the world," said Meredith Martin, assistant professor of forestry at North Carolina State University, who found that non-profit tree-planting initiatives in the tropics tend to prioritise people's subsistence needs over biodiversity or carbon storage. Over time, he said, these efforts risk reducing biodiversity in forests.

Non-profit tree-planting groups often say they plant non-native species because local communities ask them to. But a deeper commitment may tell a different story, said Susan Chomba, who oversees forest restoration and conservation in Africa for the World Resources Institute, a global non-profit research group. When given the opportunity to consider what they want to achieve on their land, farmers remember, for example, that when they had more trees, they also had streams, he said. They want their water back.

"Then you say, 'According to your traditional and local knowledge, what kind of tree species are suitable for putting water back into the ecosystem?'" continued Chomba. "They will give you a whole range of indigenous tree species."

A major obstacle is the lack of supply in local seed banks, which are often dominated by popular commercial species. Some groups overcome this problem by paying people to collect seeds from nearby forests.

Experts say that another solution is to let forests regrow on their own. If the area is scarcely degraded or is close to an existing forest, the method known as natural regeneration can be cheaper and more effective. Simply fencing off certain areas to prevent grazing often allows trees to regrow, sequestering carbon and biodiversity.

"Nature knows much more than we do,
https://www.nytimes.com/es/2022/03/21/espanol/reforestacion-pros-cons.html#:~:text=The%20reforestation%20of%20%C3%B3n%20trees%20can,badly%2C%20can%20accelerate%20the%20extinction%20of%20trees%20can,badly%2C%20can%20accelerate%20the%20extinction%C3%B3n.

V. EVOLUTION OF THE SOWING LIFE PROGRAMME

Where did this project start?

The first stage began in 2019 in Veracruz, Chiapas, Tabasco and Campeche, with the integration of 220,000 farmers to work 550,000 hectares of crops.

The second stage started in 2020 and was implemented in Colima, Guerrero, Hidalgo, Michoacán, Morelos, Oaxaca, Puebla, Quintana Roo, San Luis Potosí, Tabasco, Tamaulipas, Tlaxcala, Yucatán, Sinaloa, Durango and Chihuahua.

And by 2021, Agroforestry and Intercropped Milpa Systems with Fruit Trees are fully established on 1.75 million hectares. Between 2022 and 2023, Sembrando Vida entered into the consolidation stage in the social, productive and social finance components. 2024 will be the year of Sembrando Vida's autonomy.

Results of the Sembrando Vida Programme

Senators, farmers and civil servants highlighted the advantages and results of the Sembrando Vida programme. million hectares, which five years ago were abandoned, are now productive.

During the forum "The transformation of the Mexican countryside", Senator Leticia Peña Ocampo, from Morena, acknowledged that for decades, peasants and producers were forgotten, and that the social inequality that Mexico experienced in the 20th century "brought about the convergence of social and political leaders who sought better living conditions for the people".

He recalled that at the beginning of President Andrés Manuel López Obrador's six-year term, it was clearly defined that the focus of public policies would be the poorest groups, among them, the peasants, who feed us and, regardless of the day and weather conditions, get up very early to nurture the land.

In this regard, he said that at the beginning of 2019, the Sembrando Vida programme was implemented, which aims to contribute to the social welfare of Mexican farmers by promoting food self-sufficiency through the establishment of agroforestry systems, which combine the production of traditional crops with fruit and timber trees, in order to increase production and reduce the use of inputs and costs for the production of the plots.

Senator Gabriel García Hernández said that now Mexico can tell the world that it is possible to return to the countryside, not only to generate economic prosperity, but also to heal the planet and guarantee the sustainability of future generations.

He added that each fruit tree seed from the Sembrando Vida programme is a symbol of what the future of our country should be and a reason to return to looking at the countryside as the engine of development and the preservation of our identity and culture.

Hugo Raúl Paulín Hernández, Undersecretary for Productive Inclusion and Rural Development of the Ministry of Welfare, stressed that this is a rural development strategy that promotes collective work and self-employment in states such as Chiapas, Oaxaca, Guerrero, Puebla, Hidalgo, Morelos, Michoacán and Tlaxcala. The official indicated that it is an effective project, which is made up of 18,000 groups throughout the country, 18,000 farmer learning communities working in coordination and solidarity.In addition, she said, it has been an opportunity to recognise the work and participation of rural women in the Mexican countryside.

Paulín Hernández indicated that after six years of this investment by the Federal Government in the countryside, it is bearing fruit, a source of work and income for the farmers who participated in this project.

For this reason, he emphasised, they will propose that more hectares in the country be linked to Sembrando Vida, because it has shown that, thanks to the work of the planters and sowers, the Mexican countryside can grow and move forward. Now, he said, there is production for self-consumption and there is a surplus that generates wealth for producers. Geni González Iglesias, sower of "Tixtoca Nemeliztlik", from the municipality of Tixtla, Guerrero, said that Sembrando Vida has been transcendent and has generated social awareness and a work that has transformed the life of many farmers, and thanks to this initiative many are now self-employed and the economy has improved in many households.

VI. EXPERIENCES IN SOWING LIFE

Work was carried out with different engineers in charge of consultancy and maintenance of the different crops that were being grown in the nurseries. In the area of fertilisation, organic products were used, which was what was recommended to the producer. Some photos of some of the nurseries in the State of Mexico, from where the trees are supplied to the areas near the nurseries. We worked with the help of the scholarship holders and the engineers in charge in the area. There are agronomists, biologists and even a doctor who comes sporadically. The services provided by the scholarship holders are crop consultancy. In this area, 9 varieties of chilli (hot), maize, beans and watermelon are grown, and some products such as foliar fertilisers are made with organic products. They also teach them how to make pizzas, salt bread and sweet bread.

Photos of an ejido located in the highlands of the State of Guerrero.

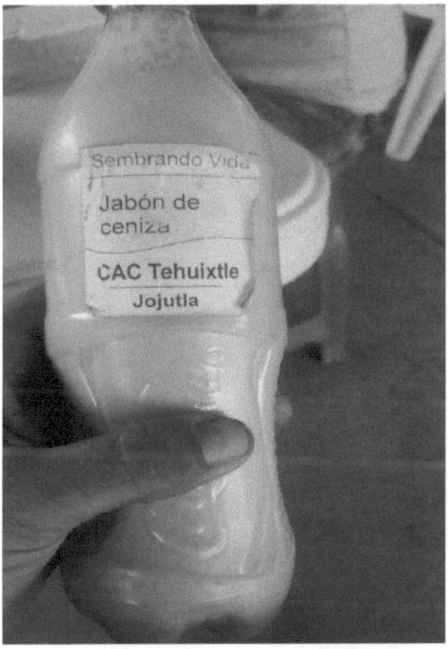
Sembrando Vida
Jabón de
ceniza
CAC Tehuixtle
Jojutla

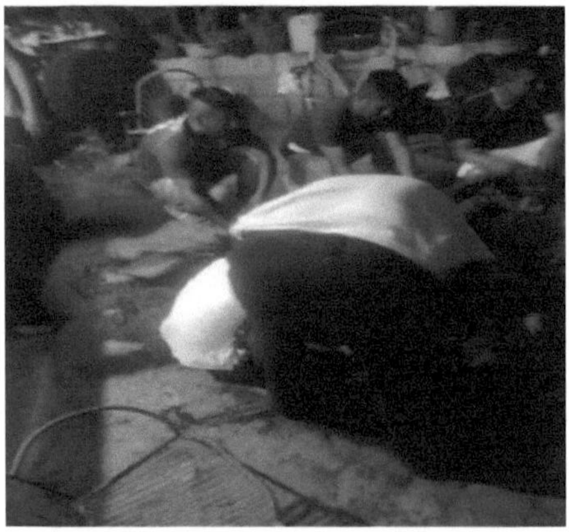

Reforestation in area plagued by Popo Park

In some areas of the State of Mexico, we visited nurseries and talked with producers and scholarship holders.

Sembrando Vida has created more than 400,000 jobs in the countryside; for its thousands of planters and seeders it is no longer necessary to leave their places of origin in search of a livelihood, they now work in their own communities.

VII. BIBLIOGRAPHY

Benefits of reforestation

https://www.google.com/search?q=importancia+and+benefits+of+the+reforestation&rlz=1C1UUXU_enMX940MX940&oq=importance+and+benefits+of+the+reforestation&g
Bethanie Walder. Society for Ecological Restoration. Global non-profit organisation.
biodiversity.https://www.iberdrola.com/sostenibilidad/que-es-reforestation#:~:text=Problems%20of%20reforestation%20and%20reforestation%20&text=We%20can%20grow%20large%20forest%20crops%20and%20position%20them%20can%20be%20damaging.
Catrin Einhorn. 21 Mar 2022 The New York Times. Get it sent to your inbox.
(FAO) 2020. The state of the world's forests
Geni González Iglesias, planter of "Tixtoca Nemeliztlik", from the municipality Robin Chazdon, Professor of Tropical Forest Restoration at the University of the Sunshine Coast, in Tixtla Australia, Guerrero (2023)
SEMBRANDO VIDA Programme. Sustainable Communities Programme. Welfare Secretariat : 06 November 2020
National Forest Programme 2020-2024

Robin Chazdon, Professor of Tropical Forest Restoration at the University of the Sunshine Coast, Australia.
University College London and the University of Edinburgh assessed national commitments to reforestation and restoration.
https://www.nytimes.com/es/2022/03/21/espanol/reforestacion-pros-cons.html#:~:text=The%20reforestation%20of%20%C3%B3n%20of%20

%C3%A1trees%20can%20be wrong

%2C%20puede%20acelerar%20la%20extinci%C3%B3n.

More Books!

yes I want morebooks!

Buy your books fast and straightforward online - at one of world's fastest growing online book stores! Environmentally sound due to Print-on-Demand technologies.

Buy your books online at
www.morebooks.shop

Kaufen Sie Ihre Bücher schnell und unkompliziert online – auf einer der am schnellsten wachsenden Buchhandelsplattformen weltweit! Dank Print-On-Demand umwelt- und ressourcenschonend produzi ert.

Bücher schneller online kaufen
www.morebooks.shop

info@omniscriptum.com
www.omniscriptum.com

OMNIScriptum

MIX
Papier aus verantwortungsvollen Quellen
Paper from responsible sources
FSC® C105338

Printed by Books on Demand GmbH, Norderstedt / Germany